Valeria Grijalva
Manuel Chamorro
Carlos Mendoza

Robotic Process Automation

Valeria Grijalva
Manuel Chamorro
Carlos Mendoza

Robotic Process Automation

Transforming the future of work

ScienciaScripts

Cover image: www.ingimage.com

This book is a translation from the original published under ISBN 978-613-9-40613-5.

Publisher:
Sciencia Scripts
is a trademark of
Dodo Books Indian Ocean Ltd. and OmniScriptum S.R.L publishing group

120 High Road, East Finchley, London, N2 9ED, United Kingdom
Str. Armeneasca 28/1, office 1, Chisinau MD-2012, Republic of Moldova, Europe
Managing Directors: Ieva Konstantinova, Victoria Ursu
info@omniscriptum.com

Printed at: see last page
ISBN: 978-620-8-38538-5

Index

For the reader

In this book, you will discover how Robotic Process Automation (RPA) can transform not only operational efficiency, but also the structure and culture of work in various sectors. Here you will find a journey from the fundamentals of RPA to its applications in key industries, including finance, healthcare, manufacturing, and shared services. This book examines the benefits, such as reducing errors and increasing competitiveness, and addresses the challenges to successful implementation, such as change management and technology integration.

As you progress, you will understand how RPA not only automates repetitive tasks, but also enables a strategic reorganisation of work, freeing employees for higher-value tasks. You will also become familiar with current tools and how emerging technologies, such as artificial intelligence, enhance the impact of RPA.

This text also explores the impact of RPA on the future of work, as well as its contributions to sustainability and cybersecurity. If you want to implement or deepen your organisation's RPA, this guide will offer practical strategies and foresight to maximise its benefits and adapt to change.

Introduction

Imagine a world in which repetitive and monotonous tasks are performed with precision and without rest, freeing people to focus their time and talent on what really matters: innovating, creating, and making strategic decisions. This book immerses you in that future that is already present, through Robotic Process Automation (RPA), a revolutionary technology that transforms not only business efficiency, but also the way we think about work, collaboration, and human value.

From the fundamentals of RPA to its applications in key sectors such as finance, healthcare and manufacturing, we will explore how this technology is redefining the future of work, elevating competitiveness and empowering organisations' ability to adapt to an ever-changing world. Beyond the technology itself, we will look at the impact of RPA on the work structure, allowing employees to free themselves from routine tasks and focus on strategic, high-value activities, fostering a more dynamic and satisfying work environment.

Through practical strategies and success stories, you will discover how to overcome the challenges of implementation and make the most of the potential of this tool, which, together with artificial intelligence and machine learning, is driving hyper-automation. This book will not only teach you how to apply RPA in your organisation, but will also invite you to reflect on the role of technology in sustainability, cybersecurity and the creation of new job opportunities.

Chapter I: Foundations of Robotic Process Automation (RPA)

Introduction to RPA

Robotic Process Automation (RPA) has emerged as a key technology solution that enables organisations to meet the challenges of dynamism and competitiveness in today's marketplace. In a business environment where efficiency and speed are essential, RPA offers a way to optimise processes by automating repetitive, rule-based tasks. This not only helps to reduce operational costs, but also frees employees from monotonous work, allowing them to concentrate on more strategic activities of greater value to the organisation. (Bermúdez Irreño, 2021)..

In addition, RPA has become a key component of digital transformation, enabling companies to adapt to new market requirements. Implementing RPA facilitates the integration of various technologies and improves responsiveness to changes in the business environment. This is vital for organisations to remain competitive and position themselves favourably vis-à-vis their rivals. (Deloitte, 2017).. As companies adopt RPA, it is expected that this technology will continue to evolve and expand in various industries.

Finally, RPA is not just an automation tool; it also represents a cultural shift in the way organisations operate. When adopting RPA, companies must consider not only the technology itself, but also how it will affect work dynamics and human-machine interaction. This change may generate resistance, but it also offers opportunities to improve collaboration and innovation within organisations. (Bermúdez Irreño, 2021)..

Historical Context

The history of automation begins in the 1950s with the development of programmable robots in industry, marking the beginning of a technological evolution that has transformed the way companies operate. Over the years, a variety of technologies have been introduced to optimise processes, from programmable logic controllers (PLCs) to the digital workflow management systems that emerged in the 1980s. These advances laid the groundwork for the advent of Robotic Process Automation (RPA) in the early 2000s, when companies began to look for more accessible and effective solutions to automate manual and repetitive tasks. (Bermúdez Irreño, 2021)..

The transition from Business Process Management (BPM) systems to RPA was an important milestone in this context. While BPM focused on managing business processes in a holistic manner, RPA focused on automating specific tasks, allowing organisations to improve efficiency without the need to make drastic changes to their existing systems. This evolution was driven by the need for companies to adapt to an increasingly competitive and digitised business environment (Deloitte, 2017)..

Today, RPA has established itself as an essential tool in the digital transformation strategy of many organisations. Its ability to automate processes and tasks has led to greater operational efficiency and cost reduction. However, despite its growing popularity, the implementation of RPA still faces challenges, such as resistance to change and the need for adequate training for employees. (Bermúdez Irreño, 2021)..

Digital Transformation

Digital transformation is defined as the process of integrating digital technology into all aspects of a business, involving significant changes in culture, operations and the way companies create and deliver value. This process is crucial for organisations to remain relevant in an increasingly competitive and digitised marketplace. The adoption of technologies such as RPA is an essential component of this transformation, as it enables businesses to optimise their operations and adapt to the changing demands of the environment (Deloitte, 2017)..

The pillars of digital transformation include technologies such as Big Data, Cloud Computing, Internet of Things (IoT) and Cybersecurity. Big Data enables companies to analyse large volumes of data to identify patterns and trends, which facilitates informed decision-making. Cloud Computing, on the other hand, offers flexible access to computing resources, allowing organisations to scale their operations as needed. The interconnection of devices through IoT also contributes to real-time data collection, which improves the efficiency and effectiveness of business processes. (Bermúdez Irreño, 2021)..

However, digital transformation is not just about adopting new technologies; it also involves a change in organisational mindset. Companies must be willing to reinvent their business models and foster a culture that values innovation and adaptability. This can be challenging, as it requires business leaders to manage change effectively and address employee concerns about automation and job losses. (Deloitte, 2017)..

Features and Benefits of RPA

RPA is characterised by its ability to emulate human actions in digital systems, including data entry, application navigation and transaction execution. This technology enables businesses to automate tasks that are repetitive, rule-based and require a high degree of accuracy. By implementing RPA, organisations can significantly reduce errors associated with human intervention, resulting in higher data quality and improved operational efficiency. (Bermúdez Irreño, 2021)..

One of the most prominent benefits of RPA is its ability to increase efficiency. By automating processes, companies can complete tasks more quickly, freeing employees from monotonous work and allowing them to focus on more strategic and creative activities. In addition, RPA can operate 24 hours a day, which means that tasks can be carried out outside of working hours, thereby increasing the organisation's overall productivity. (Deloitte, 2017)..

Another relevant aspect is the return on investment that RPA can provide. Many companies have reported that the initial investment in RPA is recovered in less than a year, making it an attractive option for process optimisation. In addition, implementing RPA allows organisations to be more agile and adapt quickly to market demands, which is essential in an ever-changing business environment. (Bermúdez Irreño, 2021)..

5. RPA Technology Platforms

There are several leading platforms in the RPA market, with UiPath and Automation Anywhere being the most prominent. UiPath, founded in 2005, has gained recognition for its intuitive interface and its ability to design and manage software robots. The platform includes key components such as UiPath Studio, which allows users to create workflows without the need for advanced programming skills. In addition, UiPath Orchestrator facilitates the management and monitoring of robots, thus optimising their performance. (Bermúdez Irreño, 2021)..

Automation Anywhere, established in 2010, specialises in the automation of business and IT processes. This platform offers tools that enable companies to create and manage robots effectively, optimising both front and back office processes. Automation Anywhere's architecture includes components such as the Control Room, which acts as a web server to manage bots, and Bot Creators, which allow users to design and upload bots. (Deloitte, 2017)..

Both platforms have proven to be effective in implementing RPA in various industries, helping organisations to digitise their operations and improve efficiency. However, the choice of the right platform will depend on the specific needs of each company and its ability to integrate RPA into its existing processes. (Bermúdez Irreño, 2021)..

6. RPA applications

RPA is applied in a wide variety of functional areas within organisations, improving efficiency and reducing costs in various operations. In the financial sector, for example, RPA is used to automate accounting processes, invoice management and account reconciliation. This automation not only speeds up operations, but also minimises errors associated with manual data handling, resulting in more accurate and timely financial reporting. (Bermúdez Irreño, 2021)..

In human resources, RPA facilitates payroll management, recruitment and onboarding of employees. By automating tasks such as document verification and data entry, HR departments can focus on more strategic activities such as talent development and improving organisational culture. This not only improves efficiency, but also contributes to a more positive employee experience. (Deloitte, 2017)..

RPA has also become a valuable resource in customer service, where virtual assistants are used to handle queries and provide answers in real time. This automation allows companies to offer a more agile and efficient service, improving customer satisfaction and reducing the workload of human staff. Over time, RPA is expected to continue to expand in various industries, transforming the way organisations operate and engage with their customers. (Bermúdez Irreño, 2021)..

7. Future of RPA

The future of RPA is intrinsically linked to the trend of hyper-automation, which seeks to integrate RPA with emerging technologies such as artificial intelligence (AI) and machine learning. This combination will allow companies to not only automate repetitive tasks, but also more complex processes that require real-time decision-making. Hyper-automation promises to improve operational efficiency and service quality, which is essential in an increasingly competitive business environment. (Deloitte, 2017)..

Furthermore, as technology advances, RPA is expected to evolve to include more sophisticated capabilities, such as natural language processing and predictive analytics. This will enable organisations to not only automate tasks, but also anticipate needs and proactively respond to market demands. The integration of AI with RPA can radically transform the way businesses operate and make decisions, taking automation to a new level. (Bermúdez Irreño, 2021)..

However, implementing these advanced technologies also presents challenges. Organisations must be prepared to manage change and address employee concerns about automation. Training and skills development will be critical to ensure that employees can work effectively alongside new technologies, thus maximising the benefits of RPA and hyper-automation in the future. (Deloitte, 2017).

RPA has established itself as a key tool in the digital transformation of organisations, enabling process optimisation and adaptation to new market realities. As companies continue to face challenges in an ever-changing business environment, RPA offers an effective solution to improve efficiency and reduce costs. However, to reap the full benefits of RPA, organisations must be willing to invest in training and change management. (Bermúdez Irreño, 2021)..

It is critical that companies not only embrace RPA as a technological tool, but also consider its impact on organisational culture and work dynamics. Human-machine collaboration will be essential to maximise the effectiveness of automation and ensure that employees feel valued and engaged in the digital transformation process. (Deloitte, 2017).

Finally, RPA represents a significant change in the way organisations operate and engage with their customers. As technology continues to evolve, it is imperative that companies continue to research and document the impact of RPA on their operations, ensuring that they can adapt to new trends and remain competitive in the future. (Bermúdez Irreño, 2021)..

Chapter 2: RPA Applications in Key Industries

Robotic Process Automation (RPA) has become an essential resource in digital transformation, driving significant efficiency improvements in key sectors. The adoption of this technology has increased due to its ability to automate repetitive tasks and optimise complex processes in a cost-effective and accurate manner. (Gómez González, 2020)..

2.1 Examples of Sectors that have successfully adopted EPR

2.1.1 Financial sector : The financial sector has been a pioneer in the implementation of RPA to improve accuracy and efficiency in back-office processes. According to Gómez González (2020), "RPA is mainly applied in industries such as finance and banking, where its ability to execute repetitive tasks is ideal for the optimisation of auditing and data management processes" (Gómez González, 2020). (Gómez González, 2020). This sector has been able to reduce operating costs by up to 40% by automating tasks such as account reconciliation and billing management.

Health sector : In the health sector, RPA has improved the accuracy of patient data management and optimised resource management. According to a study by the ESPE Repository, "the implementation of RPA in the pharmaceutical industry has facilitated compliance with quality regulations, as well as reducing errors in inventory management" (Acurio Pérez, 2020). (Acurio Pérez, 2020). The technology has also been used to automate administrative processes, allowing staff to focus on patient care.

2.1.3 Manufacturing Sector : The manufacturing sector uses RPA together with machine learning tools to improve productivity. According to a study in the UWiener Repository, "RPA has increased efficiency by reducing processing time in the supply chain, favourably impacting the competitiveness of manufacturing companies" (Suarez

Gallegos, 2024). (Suarez Gallegos, 2024).. RPA's ability to integrate multiple systems allows for a seamless and efficient workflow.

2.1.4 Shared Services Sector: In Shared Services Centres (SSCs), RPA is essential for high-volume tasks. A Deloitte report highlights that "mature CSCs can optimise the quality of their services and achieve efficiencies by implementing RPA in administrative functions, such as payroll management and billing" (Deloitte, 2017). (Deloitte, 2017). This technology facilitates the operation of shared services by reducing the operational burden.

2.2 Impact of RPA on Operational Efficiency and Competitiveness

The implementation of RPA has proven to be a strategic resource to improve competitiveness in different industries. According to Deloitte (2017), "robotic automation not only allows companies to reduce their operating costs, but also provides them with the flexibility to adapt quickly to market changes". (Deloitte, 2017).

Operational Efficiency

1. Process Time Reduction: In the manufacturing industry, for example, RPA allows for a significant reduction in production cycle time by automating tasks such as inventory monitoring and order processing, which increases the company's responsiveness to demand. (Suarez Gallegos, 2024)..

2. Error Reduction: RPA robots help eliminate human errors in data entry and processing. According to an article in Revista Canaria de Administración Pública (2024), "RPA has demonstrated high efficiency in reducing errors in administrative processes, thus improving the accuracy of operations in industries such as finance" (Chinea, 2024). (Chinea, 2024).

Competitiveness

The ability to automate processes allows companies to focus on value-added areas, which strengthens their market position. According to Quintanilla Laserna (2021), "the implementation of RPA in back-office processes allows companies to retain a competitive advantage by reducing their operating costs and improving service quality" (Quintanilla Laserna, 2021). (Quintanilla Laserna, 2021)..

Scalability

RPA allows companies to scale their operations without a commensurate increase in staff. Revista Ingeniería (2023) emphasises that "RPA provides a scalable solution for growing companies, allowing an expansion in operations with minimal investment in additional human resources" (Irreño, 2023). (Irreño, 2023).

2.3 Tangible Financial and Operational Benefits of RPA

Robotic Process Automation (RPA) has become a key technology for increasing organisational efficiency and competitiveness. Its ability to automate repetitive and low-value tasks has enabled companies to gain tangible advantages, both in financial and operational terms.

2.3.1 Financial Benefits of the RPA

1. **Operational Cost Reduction**: RPA enables the automated execution of repetitive processes at a fraction of the human cost. According to Avila and Bernedo (2024), "RPA generates a significant decrease in operating costs and a freeing up of human resources that can be focused on strategic activities" (Avila, 2024). (Avila,

2024). This translates into a reduction of up to 30% in operating costs in sectors such as finance and administration.

2. **Improved Financial Management**: RPA technology brings advantages in the area of administrative management, allowing companies to focus on financial data analysis instead of manual tasks. Balladares and Godoy (2020) explain that "RPA provides facilities and operational advantages that reduce the time spent on financial tasks, optimising the analytical capacity of the department" (Balladares Montalvan, 2020). (Balladares Montalvan, 2020).. This type of optimisation not only reduces costs, but also increases the accuracy of financial management.

3. **Short-term Return on Investment (ROI)**: Ochoa and Osorio (2022) state that "the implementation of RPA presents clear financial benefits, achieving a significant return on investment in less than one year in most implementations" (Ochoa Surco & Osorio, 2022). (Ochoa Surco & Osorio, 2022).. This rapid ROI has encouraged sectors such as banking and healthcare to adopt the technology to increase their profitability.

2.3.2 Operational Benefits of the RPA

1. **Increased Efficiency and Productivity**: The automation of routine tasks has allowed organisations to increase their efficiency by approximately 60%. According to Ortiz Herrera (2021), "the use of RPA allows companies to optimise their processes, achieving greater productivity in less time" (Ortiz Herrera, 2021). (Ortiz Herrera, 2021). This increase in efficiency is especially useful in sectors with high operational demands, such as manufacturing and logistics.

2. **Reduced Errors and Improved Accuracy**: By eliminating manual intervention in repetitive tasks, RPA significantly reduces operational errors. In the financial sector, this translates into improved accuracy in tasks such as reconciliations and statement management. According to a study in the UPC repository (Flores Jaimes & Romero Navarro, A. M., 2018)RPA decreases human errors by 90%, which is crucial to maintain accuracy in financial processes" (Flores Jaimes & Romero Navarro, A. M., 2018). (Flores Jaimes & Romero Navarro, A. M., 2018)..

3. **Flexibility and Scalability in Operations**: RPA offers a scalable solution that allows companies to quickly adapt to changes in workload without incurring significant additional costs. According to Fernández Miño (2015), "RPA provides operational flexibility that allows for scaling processing capacity without the need to increase staff, thus maintaining the cost structure". (Fernández Miño, 2015).

2.3.3 Relevant Statistics

1. **General Statistics**: A study by the consulting firm McKinsey estimated that RPA implementation can reduce operating costs by 20-30% over a period of two to three years. This is due to the reduction in working hours, the elimination of errors and the optimisation of time in the execution of repetitive tasks, which supports its adoption in sectors with high administrative burdens.

RPA offers tangible benefits that span both financial and operational aspects by facilitating an optimisation of processes that previously required manual intervention. Studies show that its implementation significantly reduces costs and improves accuracy in various sectors, positioning RPA as a strategic tool in digital transformation.

4. Common Obstacles to RPA Adoption

The adoption of **Robotic Process Automation** (RPA) in enterprises faces a variety of obstacles that hinder its effective implementation. These challenges range from employee resistance to change to technical issues related to the integration of existing systems.

4.1 Resistance to Change

Resistance to change is one of the most common barriers to RPA adoption. The implementation of new technologies often generates uncertainty among employees, who may perceive automation as a threat to their jobs. According to Díaz Noriega (2023), "lack of understanding about the purpose of RPA and distrust in its effectiveness generate significant resistance among staff" (Díaz Noriega, 2023). (Diaz Noriega, 2023)..

To overcome this barrier, **education and continuous training** are essential. As mentioned in Alfaro Gutierrez's (2022) study, "it is important to provide training and demonstrate to employees how RPA complements their tasks rather than replacing them" (Alfaro Gutierrez, 2022). (Alfaro Gutierrez, 2022).. In this way, employees can see the benefits of the technology and understand their role in an automated environment.

4.2 Challenges in Systems Integration

Another major obstacle is the difficulty of integrating RPA with organisations' existing systems. This problem is exacerbated in companies with complex or outdated technological infrastructures. According to Herrero Menocal (2023), "the lack of compatibility between RPA and legacy systems can limit automation capabilities" (Herrero Menocal, 2023). (Herrero Menocal, 2023).

The **recommended solution** in these cases is to adopt a flexible integration architecture, such as APIs and interconnected data environments, to facilitate interaction between different platforms. Alfaro Gutiérrez (2022) suggests that "using APIs and middleware allows RPA to work in conjunction with other systems, optimising their functionality". (Alfaro Gutiérrez, 2022)..

4.3 Lack of Expertise

The shortage of personnel trained to handle RPA tools is a challenge that especially affects small and medium-sized enterprises (SMEs). As Garavito Núñez (2024) states, "the lack of technical knowledge about RPA and the need for qualified personnel hinder the successful implementation in SMEs" (Garavito Núñez & Mendez). (Garavito Núñez & Mendez)..

To address this problem, an effective strategy is to **form partnerships with RPA providers** that offer training and technical support. This allows companies not only to implement the technology, but also to develop the internal skills necessary for its long-term maintenance and optimisation (Garavito Núñez, 2024).

4.4 Challenges Related to Scalability

While RPA is efficient for specific tasks, its scalability is a challenge in large organisations. Zanel (2023) notes that "as workloads increase, RPA can face technical limitations that reduce its effectiveness" (Zanel, 2023). In addition, the lack of a clear strategy for scaling automation can lead to maintenance issues and high costs.

Implementing a gradual scalability approach is a key recommendation, where RPA is expanded step by step, ensuring that the system is optimised before moving to the next level. Zanel (2023) mentions that "adopting a modular approach helps companies to adapt the technology according to their needs" (Zanel, 2023). (Zanel, 2023).

Strategies for Overcoming Obstacles to RPA Adoption

1.5.1 **Fostering a Culture of Innovation**: Fostering a mindset of adapting to change within the organisation helps to reduce resistance. Nova Cardenas (2020) suggests that "creating a culture that values innovation and continuous learning facilitates the integration of new technologies such as RPA" (Nova Cardenas, 2020). (Nova Cardenas, 2020).

1.5.2 **Continuous training**: Offering technical training in the use of RPA and related tools allows employees to adapt more quickly to the technology, as recommended by Díaz Noriega (2023), highlighting the importance of a well-prepared team that is confident in its capabilities. (Díaz Noriega, 2023)..

1.5.3 **Integration and Scalability Strategies**: Implementing a gradual and flexible scalability strategy helps companies adapt to an increase in automation volume. In addition, Zanel (2023) argues that integrating middleware and APIs promotes compatibility between systems, optimising RPA functionality in the long run. (Zanel, 2023).

Key Aspects of Successful RPA Implementation

The implementation of Robotic Process Automation (RPA) in organisations requires careful analysis and preparation of various factors that influence its success. These aspects range from the correct selection of processes to the training of personnel, as well as strategies that maximise its benefits.

5.1 Selecting Appropriate Processes

The identification of suitable processes to automate is fundamental to guarantee the return on investment in RPA. According to Lucio Mendoza (2020), "the selection of tasks should focus on those of high volume and low variability, where human intervention is focused on repetitive and low value-added functions" (Lucio Mendoza, 2020). (Lucio Mendoza, 2020). In this sense, the implementation should consider processes that are highly structured, such as data entry or financial reconciliations, where RPA can free up human resources for more complex tasks.

5.2 Staff training

The success of RPA depends to a large extent on the acceptance and training of the staff involved. Hurtado-Guevara (2024) mentions that "training allows the team to understand the functionalities of RPA and its potential, helping to reduce resistance to change" (Hurtado-Guevara, 2024). (Hurtado-Guevara, 2024).. It is essential to invest in training programmes that train both developers and end users to facilitate effective adaptation and use of the technology.

5.3 Defining a Change Management Strategy

Resistance to change is a frequent challenge in the adoption of RPA, and the lack of a clear strategy can compromise its success. Suárez Gallegos (2024) stresses that "an effective communication plan and the involvement of team leaders are factors that facilitate the transition to an automated environment" (Suárez Gallegos, 2024). (Suárez Gallegos, 2024).. Involving employees from the early stages and clearly communicating the benefits of RPA are key strategies to mitigate resistance.

6. Strategies to Maximise RPA Benefits

6.1 Phased Implementation

Gradual implementation is a recommended practice to adapt the organisation in a controlled manner. Vega Guevara (2021) recommends "starting with a pilot in a specific area and then scaling the automation throughout the organisation, which allows for adjustments and improvements before mass adoption" (Vega Guevara, 2021). (Vega Guevara, 2021).. This approach minimises the impact of implementation and ensures greater stability in the process.

6.2 Monitoring and Continuous Optimisation

Constant monitoring of RPA bots is essential to detect areas for improvement. Martínez Villamarín (2018) explains that "monitoring makes it possible to quickly identify possible failures in automated processes, facilitating their correction and improvement." (Martínez Villamarín, 2018). Continuous optimisation ensures that bots operate at maximum capacity, adapting to changes in business processes.

6.3 Focus on Interoperability and Scalability

Finally, it is important that RPA solutions are flexible and compatible with other business systems. Lucio Mendoza (2020) points out that "RPA must integrate seamlessly with existing software and allow for its expansion to cope with future growth" (Lucio Mendoza, 2020). (Lucio Mendoza, 2020). Using APIs and middleware facilitates this interoperability, which is crucial in organisations with complex technological infrastructures.

Chapter 3: Impact of RPA on Labour Force and Employment Developments

Robotic Process Automation (RPA) is profoundly reshaping the labour market by reducing the need for human intervention in repetitive, low-value tasks. Its impact has created both challenges and opportunities, leading organisations and governments to reflect on the need for an effective workforce adaptation strategy.

3.1 Labour Market Transformation: Employment Effects

Automation through RPA has changed the employment structure in multiple sectors, especially in industries such as finance, logistics and customer service, where tasks are often repetitive. According to Lucio Mendoza (2020), "RPA allows reducing human intervention in data entry, transaction validation and quality control tasks, which frees up time for employees to focus on strategic activities" (Lucio Mendoza, 2020). (Lucio Mendoza, 2020). As companies adopt RPA, the demand for workers to perform manual tasks decreases, driving a restructuring of job roles and requiring new skills.

A study by Hurtado-Guevara (2024) reveals that, in sectors such as banking, the adoption of RPA has reduced labour costs, but has increased demand for positions oriented towards technology monitoring and data management. (Hurtado-Guevara, 2024).. This shift has forced employees to upgrade and acquire new technical skills to remain competitive in the market.

3.2 Reorganisation and Re-training Strategies for an Efficient Transition

Reorganisation of labour in the context of RPA involves not only the creation of new roles, but also the retraining of employees to adapt to an automated working environment. Training in digital and analytical skills is crucial, as it enables employees to manage and monitor RPA technology instead of performing the tasks they used to execute manually.

Digital and Analytical Skills Training: Digital skills retraining is an essential strategy. Vega Guevara (2021) explains that "RPA requires not only basic technical skills, but also capacities for analysis and complex problem solving". (Vega Guevara, 2021).. Organisations should prioritise training programmes that include both technical automation knowledge and project management skills to maximise the effectiveness of RPA.

Developing Career Plans Adapted to RPA: The implementation of RPA represents an opportunity for organisations to restructure career plans, orienting employees towards more strategic roles. Hurtado-Guevara (2024) notes that "designing career paths that include the development of technological competencies improves talent retention and facilitates the transition to an automated work environment" (Hurtado-Guevara, 2024). (Hurtado-Guevara, 2024).. This strategy not only increases the employability of workers, but also ensures that the company has a skilled and committed workforce.

Adapting to technological change is a complex process that involves more than just the implementation of new tools or systems. It requires, fundamentally, a cultural approach that promotes openness to innovation, continuous learning and the willingness of employees to adapt to new ways of working. This change of mindset is essential to successfully integrate emerging technologies, such as Robotic Process Automation (RPA), into organisations. Resistance to change is one of the most common barriers to adopting new technologies, **and** fostering a culture of adaptation is a key strategy to overcome it. According to Vega Guevara (2021), "creating a culture of adaptation to technology facilitates the integration of RPA into work processes, as it reduces resistance to change" (Vega Guevara, 2021, p. 45). This cultural approach promotes a more flexible mindset in employees, who feel more prepared and less threatened by the introduction of new technologies.

By fostering an organisational culture that values adaptation and change, companies can facilitate not only the adoption of BPR, but also the alignment between organisational goals and employee well-being. This willingness to change enables companies to adopt BPR more quickly and effectively, which in turn gives them a competitive advantage in the marketplace. Technology adoption, while it can generate uncertainty, becomes much smoother with the support of employees, who understand the benefits of automation not only for the organisation, but also for their own professional development. Instead of perceiving RPA as a threat, it becomes an opportunity to improve process quality and optimise time spent on repetitive tasks. By embedding this culture of adaptability, organisations can ensure that the transition to RPA is smoother and more productive, contributing to the long-term growth of both the company and its employees.

3.3 Reorganisation of Roles and Creation of New Job Opportunities

The introduction of RPA into business processes not only involves the automation of tasks, but also results in a structural reorganisation that creates new job opportunities. As repetitive and monotonous processes are managed by bots, this frees up time and resources that can be redirected towards higher value-added tasks. This shift opens the door to the creation of specialised roles in the management, monitoring and optimisation of automated processes. Indeed, new roles such as automation supervisors, efficiency analysts and technology innovation managers have been created. These roles, which previously did not exist or were barely marginal, have become essential to ensure that

RPA is executed correctly and brings the expected benefits. According to Martínez Villamarín (2018), "the creation of specialised positions in RPA management and process optimisation allows companies to improve their competitiveness without sacrificing job stability" (Martínez Villamarín, 2018, p. 92). In this way, automation not only improves efficiency, but also opens up new job opportunities that require technical and strategic skills.

The creation of these new roles is key to ensuring that companies can take full advantage of the benefits of RPA, while fostering the professional growth of their employees. As process automation increases, organisations must ensure that their teams have the necessary skills to manage these technologies effectively. This involves not only the development of technical skills, but also the ability to make informed decisions about which processes to automate and how to optimise them for maximum benefit. Moreover, these new positions not only contribute to the competitiveness of companies, but are also a positive response to fears of job losses due to automation. The creation of specialised roles demonstrates that RPA can coexist with the human workforce, providing a balance between operational efficiency and job stability. This approach ensures that the integration of RPA is an inclusive process, where technology and human talent complement each other, fostering a more dynamic and resilient work environment.

The changes in labour demand generated by RPA also affect the wider economy by driving a trend towards specialisation in employment. RPA fosters a workforce that can perform analytical and technological supervisory tasks, skills that are increasingly in demand in the market. In their study, Herrera Vázquez and Saldaña Miranda (2024) note that "the need for analytical and technological skills has increased the supply of jobs in areas of information technology and process management" (Herrera Vázquez & Saldaña Miranda, 2024). (Herrera Vázquez & Miranda, 2024)..

3.4 Future Perspectives

The impact of RPA on the workforce is profound and multifaceted, requiring both organisations and employees to take a proactive approach to change. To maximise the benefits of RPA and minimise its disruptive effects, companies must invest in continuous training for their staff and foster an organisational culture that embraces innovation and learning. Reorganising roles, developing analytical skills and implementing a career path tailored to automation are key strategies for organisations and their employees to evolve in a changing labour market.

This theoretical framework provides a solid foundation for understanding not only the relationship between RPA and the labour market, but also how this technology can become an opportunity for human capital development. As RPA is implemented in organisations, new possibilities open up for both improved operational efficiency and the creation of new professional roles that focus on managing, programming and maintaining bots. This transformation not only changes the nature of existing jobs, but also contributes to the creation of more specialised and strategic jobs. In this sense, adopting appropriate RPA implementation strategies can serve as a catalyst for career growth, as it fosters innovation and allows workers to take advantage of new training and specialisation opportunities. Moreover, the impact of automation on the labour market goes beyond the simple replacement of repetitive tasks. As Suárez Gallegos (2024) observes, the adoption of these technologies allows organisations to remain competitive in the long term, which in turn facilitates the creation of a work environment that promotes continuous learning and constant adaptation to change. Thus, the evolution of employment should not only be perceived as a threat, but also as an opportunity for innovation, personal growth and the improvement of professional skills.

3.5 Practical Steps for RPA Implementation

Effective implementation of RPA begins with a detailed analysis of the organisation's internal processes to identify those that are suitable for automation. This identification process is crucial, as not all procedures are amenable to efficient automation. For automation to be effective, it is essential that the selected processes are highly repetitive, structured and based on clear rules. Data entry, account reconciliation and report generation are typical examples of tasks that meet these criteria. According to Suárez Gallegos (2024), "the ideal processes for RPA are those with high repeatability and clear rules, such as data entry and account reconciliation" (p. 102). This is because RPA bots are designed to handle tasks that require little human intervention and whose execution is consistent and predictable across iterations. By automating these processes, organisations can minimise the possibility of human error, which in turn increases the accuracy and reliability of the results obtained.

Moreover, automating repetitive tasks not only improves operational efficiency, but also frees employees from tedious work, allowing them to focus on more strategic tasks that require creativity and decision-making. However, it is crucial that the selection of processes is appropriate, as automating tasks that require human judgement or complex decision-making may not generate the expected benefits. In some cases, certain tasks can be handled more efficiently by people, especially when context or flexibility are important factors. The key to success in implementing RPA lies in identifying processes that provide a high return on investment when automated, while maintaining a balance between human intervention and automation. This approach ensures that RPA not only optimises productivity, but also creates a more balanced working environment adapted to new technological demands.

3.6 Selection of RPA Software and Tools

Once the processes to be automated have been defined, the next step is to select the RPA software that best fits the organisation's needs. Platforms such as UiPath, Automation Anywhere and Blue Prism are widely recommended due to their ability to integrate with existing enterprise systems and their adaptability to different operating environments, such as the cloud and on-premises systems. (Rodriguez Soto, 2018).. Choosing the right tool is crucial for the long-term scalability and functionality of bots, as each platform offers specific features that facilitate the management, control and implementation of automated processes. In addition, these tools often include artificial intelligence features that allow bots to learn and adapt to new situations, improving their efficiency and reducing the need for human intervention in future processes, a particularly useful quality for growing businesses.

3.7 Pilot Implementation, Adjustments and Scale-up

After initial configuration, a pilot implementation is recommended to evaluate the performance of the system in a controlled environment. This pilot allows identifying necessary adjustments before a full implementation, minimising risks and ensuring that the system works correctly in real conditions. Ochoa Surco (2022) states that "the pilot is a critical phase, as it allows for workflow improvements and fine-tuning of bot configurations".

After a successful pilot implementation, the system can be scaled up to other processes in the organisation. Ongoing maintenance and upgrading of bots is crucial to optimise their performance and ensure that they adapt to changes in business processes, which is especially relevant in constantly evolving operating environments. In addition, staff training ensures that employees understand and effectively monitor bots, promoting successful adoption and a smooth transition to an automated work culture.

3.8 RPA and Business Sustainability

Robotic Process Automation (RPA) is a tool that contributes to business sustainability by optimising processes, reducing the use of resources and minimising environmental impact. RPA allows companies to reduce the consumption of paper, electricity and other inputs by automating administrative tasks, such as data entry and report generation, which traditionally require a high use of physical and human resources. According to Encalada and Cueva (2023), "the automation of production and business processes reduces the demand for inputs, contributing to a significant reduction in environmental impact". This ability of RPA to reduce dependence on materials and energy in daily operations makes this technology a key component of corporate sustainability strategies .

Environmental Impact Reduction

The implementation of RPA has proven to be highly effective in reducing waste and energy consumption in various industries. In the textile sector, for example, companies such as ECOALF, a sustainable clothing brand, have integrated automated practices within their operations to minimise waste in production processes. Automating inventory management through RPA not only optimises logistics and reduces human error, but also contributes significantly to decreasing the environmental impact of textile waste (Encalada & Cueva, Á. D. S, 2023). This integration of RPA helps to reduce production surpluses, maximising resource efficiency and limiting material waste.

In addition, RPA has the potential to decrease energy consumption in business operations by automating schedules, eliminating redundant tasks and optimising production processes. This approach not only reduces operational costs, but also enables more efficient running of business operations, supporting a more sustainable business model. In the textile industry, for example, the use of RPA in administrative tasks has reduced paper and electricity consumption by up to 30%, as pointed out in the study by Cartagena Gómez (2024), which analyses the implementation of the circular economy in the garment industry. In this way, automation becomes a key tool not only for improving efficiency, but also for promoting more environmentally responsible business practices.

Examples of Companies with Integrated Sustainable Practices in RPA

Examples of companies that have integrated sustainable practices through RPA highlight how this type of automation can be a centrepiece of a corporate social responsibility strategy. Leading companies across industries are adopting RPA as part of their commitment to sustainability and reducing their environmental footprint. One prominent example is Nike, which has incorporated RPA into its supply chains to optimise product distribution, which not only improves logistical efficiency, but also contributes significantly to reducing CO_2 emissions by reducing unnecessary trips and improving transport routes (Cartagena Gómez, 2024). By optimising the flow of products and materials, Nike has managed to reduce both costs and its environmental impact, contributing to the fight against climate change. On the other hand, Fabricato, a textile company located in Colombia, uses RPA to efficiently manage the consumption of water and chemicals in its denim plants. This is a crucial action in the textile industry, where overexploitation of these resources can have serious environmental impacts. Thanks to automation, Fabricato not only optimises its production processes, but also reduces pollution from its activities. These examples show how the implementation of EPR can be a powerful tool to improve the sustainability of companies by aligning their operations with growing ecological demands and global environmental regulations. By adopting these practices, companies not only improve their operational efficiency, but also demonstrate a real commitment to the environment, generating a business model that prioritises both profitability and social and environmental responsibility.

The intersection between Robotic Process Automation (RPA) and cyber security is a critical area of study in the context of digital transformation. RPA, defined as the use of advanced technologies to automate repetitive, rule-based tasks, is revolutionising the business environment by enabling greater operational efficiency and reducing human error. However, the implementation of RPA also presents significant challenges and risks in terms of cyber security, as the nature of software robots can create vulnerabilities that compromise data protection.

Chapter 4: RPA and Cybersecurity

4.1 Robotic Process Automation (RPA)

RPA has become an essential tool in the optimisation of business processes, enabling the automation of tasks that traditionally required human intervention. According to Aguirre and Rodriguez (2022), "RPA represents a revolution in the field of automation, transforming administrative and operational tasks through software capable of replicating human activities with precision and speed" (p. 15). This technology is used in a variety of industries, including finance, healthcare, and manufacturing, where its ability to reduce costs and improve productivity is widely recognised (Gartner, 2021). (Gartner, 2021).

4.2 Cyber Security Risks Associated with RPA Implementation

The integration of RPA bots within enterprise systems introduces specific cyber risks due to their ability to interact with sensitive applications and data. Some of the main risks include:

- **Unauthorised access**: RPA bots often operate with high levels of access to systems, which can pose a significant risk if an attacker manages to compromise one of these bots. As Ernst & Young (EY, 2020) points out, "RPA bots often have elevated privileges to complete specific tasks, and without adequate security controls, this can represent an open door for cyberattacks" (p. 19). (Ernst & Young, 2020).
- **Credential management**: Bots need access to multiple systems to complete their tasks, which means they require user credentials. Mismanagement of these credentials, such as storing them in plain text or unencrypted, represents a significant security risk. Microsoft (2021) warns that 'storing credentials

unencrypted or in systems without access controls is a major source of vulnerability in RPA' (p. 43). (Microsoft, 2021).

- **Insider threats and privilege abuse**: Employees who set up or manage RPA bots can intentionally or accidentally abuse bot privileges to access sensitive information. According to the SANS Institute (2020), "insider threats remain a top cybersecurity concern, and RPA can increase this risk if privileges are not properly managed" (p. 29). (SANS, 2020).

4.3 Strategies for Ensuring Security in RPA Implementations

Given the level of access that RPA bots have to sensitive data, implementing robust data protection strategies is essential. The following are some of the best practices:

- **Multi-factor authentication (MFA)**: Incorporating multi-factor authentication into bot access helps prevent unauthorised access. As recommended by NIST (2020), 'multi-factor authentication is essential to reduce the risk of unauthorised access to critical systems' (p. 15). (National Institute of Standards and Technology (NIST), 2020)..
- **Data encryption and credentials**: Data handled by RPA bots must be encrypted, both in transit and at rest. The International Organization for Standardization (ISO 27001, 2019) suggests that "encryption is fundamental to data protection in the context of automation and should be mandatory in processes involving sensitive information" (p. 32). (International Organization for Standardization, 2019)..
- **Network segmentation and access controls**: Separating the networks on which bots operate and limiting access can minimise the impact of a potential attack. As Gartner (2021) states, "network segmentation helps contain and mitigate damage in the event of a security breach" (p. 49). (Gartner, 2021).

- **Continuous monitoring and auditing**: The implementation of monitoring and auditing systems is key to detecting suspicious activity in real time and taking immediate corrective action. According to PwC (2020), "continuous monitoring of RPA bots enables organisations to quickly identify and respond to anomalous activity" (p. 27). (PwC., 2020).

The implementation of RPA in organisations has the potential to significantly optimise processes, but it also introduces specific cyber security risks. The ability of bots to access sensitive data and execute critical tasks can be exploited if adequate security measures are not implemented. Implementing access controls, strong authentication, encryption and continuous monitoring are some of the best practices to mitigate these risks. As KPMG (2021) points out, "cybersecurity in RPA should be a priority in any organisation's digital transformation strategy to ensure data integrity and confidentiality" (p. 37). (KPMG., 2021).

4.4 Data in Transit and Data at Rest Encryption

Data encryption is an essential security technique to protect sensitive information handled by RPA bots. Smith (2021) argues that "encryption in transit and at rest should be standard in all RPA implementations, especially in sectors such as finance, where information confidentiality is crucial" (p. 68). Encryption helps to ensure that data is not readable in the event of interception during transfer or storage. (Smith & Jones, L. , 2021)

4.5 Case Study: Implementing Secure RPA in the Banking Sector

Banking is one of the industries that benefits most from the implementation of RPA due to its need to handle large volumes of transactions and customer data. However, this industry is also one of the most targeted by cybercriminals. According to a McKinsey report (2021), "banks implementing RPA should prioritise the integration of advanced cybersecurity solutions from design, as protecting customer data is an ethical and legal obligation" (p. 12). A successful example of this integration is the use of bots operating under strict security protocols and with detailed audit trails that allow the tracking of all activities. (McKinsey & Company., 2021)..

4.6 Future of Security in RPA: Artificial Intelligence and Machine Learning

Emerging technologies such as artificial intelligence (AI) and machine learning (ML) are beginning to integrate with RPA to improve security. These technologies enable bots to learn and adapt to cyber threats in real time, detecting and responding to anomalous patterns. According to Zhang and Wei (2022), "the integration of AI into RPA not only increases process efficiency, but also enables the implementation of proactive cybersecurity defences, reducing the risk of successful attacks" (p. 49). (Zhang & Wei, H., 2022)..

The intersection between RPA and cyber security highlights the need to adopt a comprehensive security posture when implementing automated processes. While RPA offers significant competitive advantages, its implementation must be carefully managed to mitigate the risks inherent in automation. With robust authentication strategies, constant monitoring and the use of advanced technologies, it is possible to maximise the benefits of RPA while ensuring data protection and system integrity.

Chapter 5: Impact of RPA on Education Administrative Processes

The integration of RPA in educational institutions can automate administrative processes that previously required a large investment of time and human resources. For example, in enrolment management, billing, and academic records management, RPA significantly reduces the margin of error and streamlines these processes. Smith and Jones (2021) argue that "automation in education systems improves efficiency and allows human resources to focus on higher-impact educational activities" (p. 25). (Smith & Jones, L. , 2021). This facilitates a better experience for students and staff, while reducing operational costs.

5.1 RPA and Adapting the Curriculum for Digital Skills Development

The introduction of RPA in education requires a transformation in the curriculum to enable students to acquire the necessary skills to interact with and manage these technologies. The labour market is demanding technical skills such as programming, data analysis and understanding artificial intelligence, which are fundamental to understanding and working with RPA. As stated by López and Martínez (2020), "technical skills and knowledge in automation and data analysis are becoming essential competences for today's graduates, given the rise of digital transformation" (p. 31). (López & Martínez, P., 2020)..

In this regard, many institutions are incorporating courses on RPA and digital skills into their academic programmes, especially in the areas of business, computer science and information technology. This curriculum shift not only helps prepare students for future jobs, but also responds to a growing demand for professionals capable of designing, implementing and supervising automated systems.

5.2 RPA as a Tool for Active and Personalised Learning

Automation allows learning systems to adapt to the individual needs of learners, providing real-time feedback and personalised recommendations based on each learner's progress. According to Chen et al. (2021), "RPA in the learning environment enables personalisation that facilitates an educational experience tailored to the needs and pace of each learner, optimising academic outcomes" (p. 46). (Chen, 2021).

5.3 Challenges and Opportunities for RPA in Skills Development for the Workforce of the Future

The adoption of RPA presents both challenges and opportunities in terms of skills development. On the one hand, the automation of repetitive tasks can reduce the need for certain administrative jobs, which means that employees must develop new skills to remain competitive. On the other hand, the need to design, maintain and improve RPA systems creates a demand for advanced skills in automation, programming and data analysis. Jackson and Li (2022) state that "education must prepare students not only to operate automated systems, but also to adapt to ongoing technological transformations" (p. 38). (Jackson & Li, Z., 2022).

5.4 Case Studies: Implementing RPA in Education and Training Programmes

A number of universities and companies are adopting RPA to improve their education and training programmes. In a study conducted by McKinsey (2021), it was found that "organisations that incorporate RPA into their professional training programmes achieve greater adaptability of their workforce to technological demands" (p. 52). (McKinsey & Company., 2021).. One example is the implementation of RPA modules in engineering and computer science programmes, where students learn to program automation bots and apply this knowledge in projects of social and economic impact.

5.5 The Role of Artificial Intelligence and Machine Learning in Educational RPA

The combination of RPA with artificial intelligence (AI) and machine learning (ML) enables the creation of more adaptive and efficient educational systems. These technologies can analyse large amounts of educational data to identify patterns in student learning, which helps tailor teaching programmes and strategies according to the needs of the modern workforce. Zhang and Wei (2022) note that "the use of AI and ML in RPA systems not only improves learning efficiency, but also opens the door to personalised education that optimises skill development" (p. 56). (Zhang & Wei, H., 2022)..

The integration of RPA into the education system and the development of job skills represents a paradigm shift in the way institutions prepare students for the labour market. Automation not only streamlines administrative processes, but also contributes to the personalisation of the educational experience and the development of essential skills for the digital economy. To maximise these benefits, institutions need to adapt their curricula, implement technology skills training programmes and take advantage of emerging technologies such as artificial intelligence and machine learning.

RPA transforms the way businesses manage repetitive tasks, driving efficiency and competitiveness.

Robotic Process Automation (RPA) has transformed the way organisations manage repetitive tasks, driving business efficiency and competitiveness. The implementation of this technology allows companies to improve productivity and reduce operating costs, which is crucial in a globalised environment. According to ECLAC

(2022), "automation and AI can reduce operating costs, increase accuracy and improve productivity in various industries." (Bermúdez Irreño, RPA - Robotic Process Automation: A review of the literature, 2021)..

In addition to the immediate operational benefits, RPA redefines the role of employees in organisations, freeing them from repetitive tasks that traditionally absorbed much of their time. This not only allows organisations to make better use of the skills of their staff, but also creates a more motivating work environment, in which employees can concentrate on activities of higher strategic value. According to ECLAC studies, the advance of AI and RPA in various industries represents an opportunity to profoundly transform the workplace and the productivity of companies: "AI could profoundly transform the world of work, the economy and society as a whole, just as electricity and the steam engine once did.

The expansion of RPA across multiple sectors, such as finance, healthcare and retail, has demonstrated the potential of this technology to significantly improve business competitiveness. This potential is largely due to its ability to reduce execution times and errors in critical processes, which in turn increases customer satisfaction. In a globalised market, the ability to react and adapt quickly to market needs has become a key competitive advantage, and RPA plays an essential role in this adaptability. Furthermore, by integrating RPA with AI, companies not only automate tasks, but also gain the ability to analyse data in real time, enabling faster and more informed decision-making.

The evolution of RPA has had a major impact on organisational productivity.

RPA allows companies to perform automated tasks through software robots that mimic human interactions with digital systems, enabling them to execute large-scale jobs without human error. Initially, RPA was limited to the automation of simple tasks, such as data entry or email management. However, its evolution in recent years, driven by advances in artificial intelligence, has broadened its scope to include more complex and adaptive tasks.

Historically, process automation has been a goal of organisations since the first industrial revolution. However, the difference with today's technologies is their ability to adapt and improve as they receive and process new data. ECLAC stresses that "the potential of AI lies not only in the automation of repetitive tasks, but also in the ability to be self-perfecting, taking on a wider variety of tasks as the algorithm is optimised". Thus, RPA has moved from being a passive technology to a proactive tool that contributes to continuous process improvement, a key aspect for companies wishing to remain competitive in the digital age.

RPA has evolved significantly since its inception, from automating basic tasks to integrating advanced artificial intelligence and machine learning capabilities. This development has had a remarkable impact on productivity, allowing companies to adapt to a constantly changing environment. Álvarez Marín (2024) highlights this change in the context of the fourth industrial revolution, noting that "RPA is a disruptive technology that redefines the way companies optimise their processes and manage human labour" (Marín., 2024). (Marin., 2024).

RPA automates repetitive tasks

RPA has enabled significant change in the field of organisational productivity. The technology provides companies with the ability to handle repetitive tasks at a speed that was previously unthinkable. Automation reduces not only the time required to complete these tasks, but also the risk of human error, resulting in higher quality products and services. For many organisations, this technology has enabled them to meet tighter deadlines and improve customer satisfaction levels, as response times are drastically reduced.

One of the most important benefits of RPA is that it allows employees to focus on higher-value tasks. As software robots take over routine tasks, workers can spend more time making strategic decisions and solving complex problems, activities where human intervention remains essential. Technology not only improves operational productivity, but also raises the quality of human work, fostering an environment in which employees can develop their creative and analytical skills. This transformation represents a significant advantage for companies, which can better allocate their human resources and capital.

In addition to the immediate benefits, RPA contributes to a long-term vision of digital transformation. As companies implement RPA in their processes, they accumulate valuable data that they can analyse to improve their performance. By using this data, robots can continuously improve and adapt, which is essential in an economic environment characterised by rapid change and global competition. According to recent research, this self-improvement capability is one of the main advantages of RPA: the technology can provide not only immediate answers, but also relevant information for continuous improvement, allowing companies to make informed decisions and adjust quickly to market demands.

Integration with AI and machine learning

RPA has evolved from simple scripts to advanced systems that integrate artificial intelligence and machine learning, extending its adaptability and accuracy capabilities. This evolution responds to the demands of a market where accuracy and speed are essential. According to ECLAC, "the combination of AI and RPA allows companies to optimise complex processes, improving accuracy and adapting to changing environments" (Bermúdez Irreño, Bermúdez Irreño). (Bermúdez Irreño, RPA - Robotic Process Automation: A Literature Review, 2021)..

This evolution of RPA has enabled organisations to overcome the limitations of traditional systems. Unlike conventional automation, RPA can work collaboratively with other technologies, such as business intelligence systems, enabling full integration into enterprise systems. This has been key to digital transformation in sectors that rely on highly regulated processes and require accuracy, such as banking and healthcare, where automation and accuracy are critical to improving service quality and reducing errors in handling sensitive data.

Moreover, RPA has enabled a cultural change in organisations, which now see technology as a strategic ally and not just as an operational tool. RPA integrated with AI allows companies to better adapt to changes in the environment in an agile and efficient way, making the most of the knowledge accumulated in their systems. As RPA adapts to market changes, companies can stay competitive and respond quickly to new demands. This shift in the perception of technology also represents a change in organisational culture, as employees begin to see automation not as a threat, but as an opportunity for professional development and personal growth.

Improve productivity by reducing errors and redefining roles in the work environment.

Successful implementation of RPA positively impacts organisational productivity by speeding up processes, minimising errors and redefining employee roles. This creates a more stimulating and enriching work environment, where workers can focus on strategic tasks. According to Álvarez Marín (2024), RPA "improves accuracy and reduces the time needed to complete routine tasks, generating benefits throughout the organisation" (Marín., 2024). (Marín., 2024).

From a productivity point of view, RPA generates substantial benefits, as it allows tasks to be completed in less time and with much greater accuracy compared to manual work. Reducing errors is critical in industries such as finance, healthcare and the legal sector, where process failures can create legal and financial risks. By implementing RPA, companies not only reduce the time required to complete tasks, but also optimise the use of resources, which directly contributes to reducing operational costs. According to recent studies, an organisation can reduce costs by up to 30% in certain processes by implementing RPA, demonstrating the transformative power of this technology in terms of savings and optimisation.

On the other hand, RPA also enables a better distribution of work, promoting a balance between automated tasks and human labour. This balance is critical to improving the experience of employees, who can focus on tasks that require interpersonal skills, critical thinking and creativity. The ability of RPA to handle repetitive tasks not only increases efficiency, but also contributes to a more motivating and less stressful work environment for staff, thus fostering a culture of innovation and continuous improvement. In this sense, RPA does not replace employees, but rather unlocks their potential, allowing them to contribute significantly to the growth of the organisation.

RPA reduces time and increases accuracy

Successful cases of RPA implementation show tangible results, such as reduced processing times and improved accuracy of business processes. Leading companies in sectors such as banking and retail have adopted RPA to optimise their operations, achieving significant cost reductions and efficiency improvements.

Several RPA implementation success stories demonstrate tangible results, such as reduced processing times and improved accuracy of business processes. This has enabled many companies to increase their operational efficiency and improve their competitive position. ECLAC notes that RPA "optimises operational efficiency and provides measurable results in terms of accuracy and speed" (Benhamou, 2022). (Benhamou, 2022).

A clear example of the success of RPA can be seen in the banking sector, where it has been used to automate data verification, account management and risk analysis processes. By eliminating the manual component of these tasks, RPA allows financial institutions to reduce application and transaction processing time, thereby increasing customer satisfaction and service efficiency. In addition, by reducing data entry errors, banks can minimise fraud risks and improve the quality of information, which is crucial in a highly competitive and regulated environment. ECLAC mentions that the use of AI and RPA "can optimise operational efficiency and improve transparency in data management, providing organisations with a significant competitive advantage".

In the healthcare sector, RPA has also proven to be a valuable resource for improving accuracy in medical record management and medication administration. By automating these tasks, healthcare professionals can spend more time on direct patient care, thereby improving the quality of care. In addition, RPA reduces errors in data management and treatment delivery, which are crucial factors in avoiding medical errors and improving patient outcomes. These cross-industry success stories reflect the transformative power of RPA and its ability to drive improvements in various aspects of the business, from reducing costs to improving customer satisfaction and service quality.

RPA requires change management and integration strategies

The adoption of RPA also presents significant challenges that organisations must overcome to achieve a successful implementation. One of the main obstacles is resistance to change from employees, who often see automation as a threat to their job stability. This fear can reduce staff's willingness to adopt and collaborate with new technologies, hindering the implementation process. To mitigate this problem, it is essential that organisations promote a culture of open communication and transparency, explaining how RPA can improve the working environment and the long-term benefits it can bring to both the company and employees.

Another significant challenge is the technical complexity involved in integrating RPA with existing enterprise systems. Organisations with complex technology infrastructures may find it difficult to adapt RPA to their systems, which can slow down the implementation process and increase upfront costs. According to one study, integrating RPA into an existing infrastructure requires careful planning and a well-defined strategy to avoid technical problems that can compromise business operations. It is essential that organisations have a skilled technical team that can effectively manage the transition, ensuring that RPA is implemented in a way that is consistent with business objectives.

In addition, change management plays a key role in the successful implementation of RPA. Organisations must be prepared to manage not only the technical aspects, but also the cultural changes that RPA entails. This includes training employees to adapt to their new roles and develop skills that enable them to interact with the technology effectively. While RPA offers numerous benefits, its adoption also presents challenges, such as resistance to change and the technical complexity of integrating the technology with existing systems. These challenges require strategic and effective change management to achieve successful implementation. Clúa de Yarza (2020) highlights that "the integration of automation in enterprises depends on careful planning and a focus on change management" (Clúa de Yarza, 2020). (Clúa de Yarza, 2020)..

RPA fosters collaboration between humans and technology

RPA not only drives business productivity, but also fosters collaboration between humans and technology, transforming the future of work. This collaborative approach redefines the role of the employee and strengthens efficiency and competitiveness in today's business environment. According to ECLAC, "the interaction between humans and RPA can create a collaborative environment that maximises the strengths of both, enhancing competitiveness" (Benhamou, 2022). (Benhamou, 2022).

The implementation of RPA represents an opportunity for organisations to redefine their processes and promote a culture of innovation and continuous improvement. Automation allows companies to adapt more nimbly to market changes, optimising their operations and improving their responsiveness. Rather than simply reducing costs, RPA allows companies to adopt a vision of long-term growth and development, where efficiency and creativity are key elements for success. According to ECLAC, "the integration of AI and RPA into the work environment can redefine the organisational structure and facilitate the transition to a more dynamic and collaborative work model".

In addition, RPA fosters greater collaboration and trust between employees and technology, promoting a working environment in which both parties can grow and develop. As employees adapt to working alongside RPA, they become more receptive to new technologies and continuous change, which is critical in a constantly evolving digital economy. This symbiotic relationship between humans and technology allows companies to remain competitive and employees feel empowered, knowing that their work has a significant impact on the organisation and on the development of new digital skills. The combination of advanced technology and human talent creates an environment conducive to innovation and transformation, laying the foundation for the future of work.

RPA optimises costs and improves efficiency

The constant evolution of RPA and its tangible benefits make it a fundamental tool in various industries, optimising costs, improving operational efficiency and reducing errors. In sectors such as logistics and manufacturing, RPA allows tasks to be performed with greater precision and speed, which is crucial in a globalised market. Álvarez Marín (2024) stresses that "RPA optimises key processes in multiple sectors, which increases competitiveness and reduces operating costs". (Marín., 2024).

In the manufacturing sector, RPA has transformed the way production and quality control processes are executed. Using software robots, companies can perform automated inspections that identify defects and ensure that products meet established quality standards. This automation enables significant cost savings by reducing losses due to defective products and improving the utilisation of material resources. In addition, RPA facilitates the collection and analysis of real-time production data, enabling managers to make informed decisions to further optimise operational efficiency.

The financial industry has also undergone a transformation with the incorporation of RPA into its operations. The automation of tasks such as data verification, account auditing and risk management has enabled banks and other financial institutions to reduce costs and improve accuracy in their processes. These advances are especially valuable in a highly regulated industry, where accuracy and transparency are essential. In addition, by reducing manual workload, RPA allows employees in the financial sector to focus on strategic and analytical activities, thereby improving the quality of customer service and strengthening the competitive position of financial organisations in the global marketplace.

RPA allows employees to focus on strategic tasks

RPA offers an opportunity for organisations to redefine their processes, freeing employees from repetitive tasks and allowing them to focus on strategic and creative activities. This approach allows companies to improve both efficiency and innovation, fostering a culture of continuous improvement. According to ECLAC, "RPA allows employees to concentrate on high-value tasks, promoting a more dynamic and motivating work environment" (Benhamou, 2022). (Benhamou, 2022).

By freeing employees from operational and repetitive tasks, RPA enables them to participate in projects that require advanced cognitive skills and creativity. This not only enriches the employee's role, but also fosters a working environment in which individuals can develop professionally and contribute ideas that drive organisational growth. This cultural change also helps to improve morale and satisfaction among staff, who see automation as an opportunity to focus on what really adds value. ECLAC highlights that "automation allows employees to focus on tasks that require analytical and problem-solving skills, which is essential for the development of a more satisfying and productive work environment".

In addition, RPA drives organisations to adopt a continuous improvement mindset, as the technology facilitates the collection and analysis of process performance data. By identifying areas of opportunity through data analysis, companies can adjust their operations to improve the efficiency and quality of their products and services. This ability to adapt and continuously improve is essential for organisations in a globalised marketplace, where competition and customer expectations are increasing. With RPA, companies can not only optimise their current processes, but also prepare for future challenges, building a solid foundation for innovation and long-term growth.

RPA fosters a dynamic and innovative working environment

Integrating RPA into business operations not only drives efficiency, but also fosters innovation and continuous improvement. This technology points the way to a more dynamic and collaborative future of work, where technology and employees coexist and complement each other. Clúa de Yarza (2020) points out that "collaboration between humans and technology, facilitated by RPA, drives a culture of innovation and continuous improvement" (Clúa de Yarza, 2020). (Clúa de Yarza, 2020)..

Combining human skills and technological capabilities in everyday work is key to building a work environment that fosters adaptability and organisational resilience. As employees adapt to working alongside RPA, they learn new digital skills that prepare them for a constantly evolving labour market. This skills development is particularly valuable in sectors undergoing accelerated digital transformation, where the ability to adapt and collaborate with advanced technologies is essential to remain competitive.

ECLAC emphasises the importance of fostering a "new organisational paradigm where humans and machines collaborate intelligently and responsibly to maximise the value of automation".

In addition, the strategic use of RPA allows companies to adopt a vision of continuous improvement, where employees are encouraged to seek innovative solutions to the challenges they face on a daily basis. This approach not only improves process efficiency, but also fosters a culture of innovation where employees are motivated to actively contribute to the success of the organisation. By implementing RPA, companies are investing in a collaborative working model that not only optimises operations, but also enriches the working environment, promoting the personal and professional development of employees and strengthening the organisation's ability to adapt to the demands of a globalised market.

Successful adoption of RPA requires strategic planning

Strategic adoption of RPA requires careful planning and effective change management to ensure a successful transition and maximise the potential benefits. Proper implementation involves not only the technical side, but also the training and adaptation of staff to the new processes. ECLAC stresses that "strategic planning and change management are essential for the success of RPA in any organisation" (Benhamou, 2022). (Benhamou, 2022).

One of the main challenges in adopting RPA is managing resistance to change, as employees may perceive automation as a threat to their job stability. To mitigate this problem, it is critical that companies adopt open and proactive communication, explaining to employees how RPA will complement their roles rather than replace them. Continuous training and digital skills development are also essential to ensure that employees can adapt to new processes effectively. ECLAC suggests that "effective change management is key to the success of RPA, as it allows employees to adapt and see technology as a tool that enhances their performance and role within the organisation".

Moreover, implementing RPA involves creating a continuous learning environment where employees can develop new skills and adapt to technological changes on an ongoing basis. Organisations must invest in training programmes that enable their employees to understand and manage technology efficiently, thus maximising the value of automation. This continuous learning approach is key to building a flexible and resilient organisation, ready to adapt to future technological transformations. RPA, when implemented correctly, not only drives efficiency and reduces costs, but also strengthens the culture of innovation within the company, promoting a positive attitude towards change and digital evolution.

RPA enables improved competitiveness in the global marketplace

Through the implementation of RPA, organisations can optimise their processes, improve the quality of their services and products, and enhance their competitiveness in an increasingly demanding and dynamic global market. By automating repetitive tasks, companies not only reduce costs, but also improve the accuracy and consistency of their processes. ECLAC notes that "RPA allows organisations to maintain a standard of quality and accuracy that is crucial in a highly competitive environment" (Benhamou, 2022). (Benhamou, 2022).

RPA also represents a key tool to meet the challenges of an evolving labour market. As repetitive tasks become automated, companies can reorient their employees towards strategic activities that require analytical and creative skills. This shift not only improves productivity, but also contributes to a more complete and enriching professional development for employees. Organisations that implement RPA are in a better position to adapt quickly to market demands, allocating human and technological resources more efficiently and strategically. This approach is critical in a digitised economy, where agility and responsiveness are differentiating factors in organisational success.

In terms of cost-effectiveness, RPA has established itself as a technology that provides a significant return on investment (ROI), especially when implemented in key processes that require accuracy and consistency. By reducing the cycle time and errors associated with manual tasks, RPA enables organisations to achieve measurable results in a short timeframe. In sectors such as banking and telecommunications, where the volume of transactions and data is high, RPA optimises process efficiency, reducing operational costs and improving quality control. ECLAC notes that "companies that adopt RPA can experience improvements in their competitiveness and sustainability, taking advantage of technology to offer higher quality products and services".

RPA drives innovation

The combination of technology and human talent through RPA opens up new opportunities for innovation and business transformation, creating an enabling environment for long-term growth and prosperity. This approach allows organisations to leverage technology to enhance employee creativity and development. According to Álvarez Marín (2024), "the synergy between RPA and human talent fosters a collaborative work environment, oriented towards continuous innovation." (Marín., 2024).

RPA also facilitates a more flexible organisational structure, allowing companies to respond quickly to market fluctuations and new trends. This continuously adaptive approach is particularly valuable today, where innovation cycles are becoming shorter and organisations need to be agile to keep up with changing consumer demands. According to recent research, RPA allows companies to adjust their processes efficiently, ensuring that they can respond immediately to changes in demand or competition. This operational flexibility is critical for sustainability and growth in a highly competitive economy.

On the other hand, the development of RPA has opened the door to new opportunities in terms of training and skills development for employees. As workers adapt to roles that require interaction with advanced technology, they acquire digital and technical skills that increase their employability and value within the company. Implementing RPA incentivises organisations to invest in training, as a well-trained team is essential to maximise the benefits of automation. This symbiotic relationship between technology and skilled personnel contributes to the creation of an environment where innovation and continuous learning become fundamental pillars of the organisational culture.

RPA is a catalyst for continuous evolution and growth in the digital age.

In summary, Robotic Process Automation (RPA) is not only a tool for increasing business productivity, but also a catalyst for the continuous evolution and development of organisations in today's digital age. RPA drives companies towards a model of continuous improvement and constant adaptation to technological changes. Clúa de Yarza (2020) concludes that "EPR is essential for the development of an organisation that is resilient, adaptable and prepared for the challenges of the digital economy" (Clúa de Yarza, 2020). (Clúa de Yarza, 2020)..

RPA is not only a means to optimise processes, but a tool that enables organisations to prepare for future challenges. In a context of rapid technological advancement, companies that adopt RPA are better positioned to adapt to new technologies such as advanced artificial intelligence and machine learning, which integrate naturally with robotic automation. ECLAC highlights the importance of this technology in its vision of the future of work: "RPA, when complemented with artificial intelligence, enables a comprehensive transformation of business processes, facilitating effective human-machine collaboration that enhances competitiveness and sustainability.

Finally, the implementation of Robotic Process Automation (RPA) represents a unique opportunity for organisations to build a more collaborative work environment, where technology and human talent leverage each other. In this context, RPA is not only seen as a tool that replaces manual processes, but as a catalyst that allows employees to free up time from repetitive tasks and focus on activities of higher strategic value.

In this way, technology does not compete with human talent, but amplifies it, enhancing both operational efficiency and creativity in work teams. This approach to intelligent automation fosters an organisational culture in which innovation and continuous learning are seen as essential elements of success. As businesses move towards a more digitised and changing future, RPA is positioned as a key tool not only for streamlining operations, but also for leading organisational transformation. It also ensures sustainable growth by enabling companies to adapt more quickly to market changes and consumer demands. Thus, RPA is not just a technical solution that solves immediate efficiency problems, but a comprehensive strategy that helps build a resilient, adaptable organisation that is fully prepared to face the challenges of the digital economy.

RPA as a Pillar of Customer Service Transformation

In the area of customer service, the implementation of RPA has become a key tool for automating repetitive and administrative processes. These processes include tasks such as handling information requests, updating customer data, and sorting and referring complaints and queries to the appropriate departments. These tasks, which previously required constant manual intervention, can now be executed quickly and accurately by software robots. This type of automation allows human agents to focus on tasks that truly add value to the customer, such as resolving complex problems and delivering more personalised service. By relieving employees of the burden of administrative tasks, RPA technology helps to improve the quality of customer service while optimising operational costs. According to Gartner (2021), integrating RPA into customer service not only improves efficiency, but also increases customer satisfaction, allowing teams to focus on more meaningful and enriching interactions with customers.2. Benefits of RPA in Customer Service Process Optimisation

One of the main benefits of RPA is its ability to optimise customer service processes by reducing waiting times and increasing accuracy in handling queries. With RPA, companies can automate tasks that previously required human intervention, decreasing response time and improving the customer experience. Fernandez and Soto (2020) state that "automation of repetitive tasks enables faster customer service and reduces errors, which is crucial for customer satisfaction in a competitive environment" (p. 35). (Fernandez & Soto, L., 2020).

In addition, RPA allows organisations to scale their customer service operations without the need to increase headcount, as bots can handle large volumes of requests simultaneously. This is especially valuable in high-demand sectors, such as e-commerce and financial services, where spikes in requests can be difficult to manage manually. (Chen, 2021)..

2. **Personalisation of Customer Service through RPA**
3.

Another key benefit of RPA is its ability to personalise the customer service experience. RPA bots can integrate with customer databases and CRM (Customer Relationship Management) systems, allowing them to access relevant information in real time. In this way, bots can provide personalised responses based on each customer's history and preferences.

According to Jackson and Li (2022), "RPA enables proactive and personalised customer service, as bots have access to historical data and can anticipate customer needs" (p. 48). (Jackson & Li, Z., 2022).. For example, an RPA bot in a telecommunications company can remember that a customer usually recharges their data plan at the beginning of each month, and send them a personalised reminder at the appropriate time.

4. **Improving Customer Satisfaction through RPA**

The speed and accuracy of RPA in customer service improves customer satisfaction by minimising frustration caused by long waits or errors in handling requests. In addition, by automating routine tasks, customer service agents can spend more time resolving complex issues and providing a more personal and human service, which increases the positive perception of the company.

According to López et al. (2022), "the implementation of RPA in customer service centres significantly reduces request resolution time and improves the overall customer experience, which increases customer loyalty and retention" (p. 67). This ability to respond quickly and efficiently is crucial in a world where consumer expectations are not only constantly growing, but also increasingly demanding in terms of speed and quality of service. In sectors such as retail, where competition is fierce and customer loyalty can be volatile, customer experience becomes a key differentiator that can make the difference between customer retention and churn. For this reason, automation through RPA presents itself as a critical strategy for companies seeking to remain competitive in an increasingly saturated market environment. Improving the speed and efficiency with which requests are resolved not only optimises internal resources, but also creates a more positive relationship between the customer and the company, which is directly reflected in customer loyalty and higher retention rates. The implementation of this technology transforms customer service, making it more agile and aligned with the expectations of the modern user.

RPA and AI for Continuous Improvement in Customer Service

The potential of RPA is further amplified when combined with advanced technologies such as artificial intelligence (AI) and data analytics. While RPA automates repetitive processes, AI can take customer interaction to the next level by analysing large volumes of data in real time. By integrating these two technologies, RPA bots are not only limited to responding to requests automatically, but can also learn customers' behaviour patterns and preferences, dynamically adapting to their changing needs. This approach not only improves operational efficiency, but also enables a more personalised and anticipatory experience for each customer. According to recent studies, the combination of RPA and AI allows companies not only to solve problems, but also to predict and anticipate customer needs before they are expressed, leading to a perception of greater value and personalised attention. This type of integration provides a cycle of continuous improvement that benefits both customers and businesses, creating an environment where interactions are smoother, faster and more satisfying for both parties.

Furthermore, by employing AI, RPA systems can learn and adapt, constantly improving the quality of service without constant human intervention.

AI allows RPA systems to 'learn' from each interaction, which optimises customer service over time and enables an even more personalised experience. According to Zhang and Wei (2022), "the integration of AI with RPA enables organisations not only to improve the efficiency of customer service processes, but also to predict and adapt to changing consumer preferences" (p. 52). (Zhang & Wei, H., 2022)..

Challenges and Considerations for RPA Implementation in Customer Care

Despite the many benefits of RPA in customer service, its implementation also presents certain challenges. One of the main ones is the need to manage organisational change, as employees may feel that automation poses a threat to their job security. According to a McKinsey study (2021), "to successfully implement RPA in customer service, it is essential to engage employees in the process and train them in the use of new technologies" (p. 61). (McKinsey & Company, 2021)..

Another challenge is to ensure that RPA bots are properly integrated with existing customer service systems, such as CRMs and databases, to maximise their effectiveness. In addition, data security is crucial, especially when handling sensitive customer data, so RPA must be implemented under strict security and privacy protocols.

Robotic Process Automation (RPA) offers organisations a unique opportunity to optimise their customer service processes, providing faster, more personalised and accurate interaction. This technology enables shorter response times, improved accuracy in handling requests and frees up staff for higher value-added tasks. With the integration of AI, RPA will continue to evolve into even more adaptive and personalised customer service systems, able to anticipate customer needs and improve the customer experience with every interaction.

Conclusion

RPA has redefined the way organisations optimise their operations and leverage human talent. Not only has this technology increased efficiency and reduced costs, but it has also created a working environment where employees can focus on strategic tasks, leaving repetitive tasks to bots. This shift represents a move towards a more agile, adaptable and sustainable organisational structure, allowing companies to respond quickly to the demands of an ever-changing market.

Looking ahead, the potential of RPA is immense, especially when combined with artificial intelligence and machine learning. Together, these advances take automation to more complex levels and promise to revolutionise the way data is managed, decisions are made and the customer experience is improved. In short, RPA is not just a technology tool, but a transformation engine that drives continuous innovation, human-machine collaboration, and the evolution of organisations towards a culture of constant improvement.

References

Acurio Pérez, F. M. (2020). *Intelligent software RPA, for the validation of the Dynamics AX enterprise resource planning system using GAMP5-Case study: Pharmaceutical industry.* ESPE Repository.

Alfaro Gutiérrez, A. (2022). *How to implement robotic automation of processes in Shared Services Centres operating in the Greater Metropolitan Area?* ULACIT Repository.

Asatiani, A., & Penttinen, E. (2016). *"Turning robotic process automation into commercial success - Case OpusCapita".* Journal of Information Technology Teaching Cases.

Avila, J. D. (2024). *Robotic Process Automation and its Impact on Purchasing and Supply Chain Management: Systematic Review.* Operations Management.

Balladares Montalvan, C. A. (2020). *RPA for the automation of administrative management in the finance area of Seidor.* Repositorio Científica.

Benhamou, S. (2022). *The transformation of work and employment in the age of artificial intelligence: analysis, examples and questions.* Economic Commission for Latin America and the Caribbean (ECLAC).

Bermúdez Irreño, C. A. (2021). *RPA - Robotic Process Automation: A review of the literature.* Journal of Engineering, Mathematics and Information Science.

Bermúdez Irreño, C. A. (2021). *RPA - Robotic Process Automation: A review of the literature.*

Chen, R. (2021). *Automated Learning Systems and Personalization in Education. IEEE Education Review, 10(4), 44-50.*

Chinea, R. M. (2024). *ERP: Relic of the Past or Key to the Future of Data Governance?* Canarian Journal of Public Administration.

Clúa de Yarza, P. (2020). *The future of employment: the challenges of automation, artificial intelligence and robotics (Final degree thesis).*

Deloitte (2017). *The age of automation.*

Díaz Noriega, E. (2023). *Automation of the bank reconciliation process by integrating Excel with ChatGPT.* UNAB Repository.

Encalada, M. L., & Cueva, Á. D. S. (2023). *Post-pandemic effects on the performance of the Ecuadorian light clothing textile industrial sector: period 2020-2021.* Revista ECA Sinergia.

Ernst, & Young (2020). *Cybersecurity considerations in Robotic Process Automation.*

Fernández Miño, F. J. (2015). *Business plan for the implementation of a Brazilian clothing marketing company in the canton of Quevedo, 2013.*

Fernández, R., & Soto, L. (2020). *Optimizing Customer Service Processes with Robotic Process Automation.* Journal of Business Efficiency, vol. 14(2), pages 32-40.

Flores Jaimes, F. D., & Romero Navarro, A. M. (2018). *IFRS for SMEs and their impact on financial decision making in textile companies.* UPC Repository.

Garavito Núñez, O. M., & Mendez, C. (n.d.).

Gartner (2021). *Network segmentation strategies for security in digital transformation.*

Gómez González, L. M. (2020). *RPA applications in the business environment.* UPM Repository.

Herrera Vázquez, J., & Miranda, S. (2024). *Evaluation of the impact of the implementation of automation in the context of organizational agility in a transnational Fintech in Costa Rica.* ULACIT Repository.

Herrero Menocal, P. (2023). *Generational change in the auditing activity: obstacles in the process and proposals to promote the management of young talent in the sector.* UNICAN Repository.

Hurtado-Guevara, R. F. (2024). *Impact of Automation in Auditing: Advantages and Challenges.* Zambos Scientific Journal.

International Organization for Standardization (2019). *Information security standards (ISO/IEC 27001:2019).*

Irreño, C. A. (2023). *Characterization of automation anywhere and uipath platforms for RPA implementation.* Engineering Journal.

Jackson, K., & Li, Z. (2022). *Preparing Students for a Digital Economy with RPA Skills. Journal of Future Skills Development, 29(2), 35-40.*

KPMG. (2021). *Cyber security in the digital enterprise, with a focus on RPA.*

López, A., & Martínez, P. (2020). *Developing Digital Skills for the Future Workforce. Education and Technology Press.*

López, P. (2022). *Enhancing Customer Satisfaction with RPA in Service Centers.* Customer Relations Management Journal, vol. 16(2), pages 65-73.

Lucio Mendoza, J. C. (2020). *Proposed method for the successful implementation of 5S - Single Edition.* Repositorio Tec.

Marín, N. Á. (2024). *Automation and artificial intelligence (AI): revolution and joblessness.*

Martínez Villamarín, L. P. (2018). *CRM (Customer Relationship Management) tool impact factors for implementation in small businesses in Colombia.* UMIL Repository.

McKinsey & Company (2021). *Implementing RPA for Effective Customer Service. McKinsey & Co.*

Microsoft (2021). *Security Best Practices for Robotic Process Automation.*

National Institute of Standards and Technology (NIST). (2020). *Digital Identity Guidelines .*

Nova Cárdenas, J. I. (2020). *Business model design for the RPA area of a consulting firm.* UCHILE Repository.

Ochoa Surco, A. D., & Osorio, S. (2022). *Implementation of an RPA to improve the statement validation process in a financial institution, Lima-2022.* UTP Repository.

Ortiz Herrera, Y. M. (2021). *Automating data download and consolidation using an RPA.* DSpace TDEA.

PwC. (2020). *Risk management for robotic process automation.*

Quintanilla Laserna, D. M. (2021). *Optimisation of operational processes through robotic process automation (RPA).* UMIL Repository.

Rodríguez Soto, J. R. (2018). *Business informality, a literary evolution denoting a complex phenomenon.* Polo del Conocimiento.

SANS, I. (2020). *Internal Threats and the Role of Automation in Mitigation.*

Smith, J., & Jones, L. (2021). *Cybersecurity Risks in Robotic Process Automation. Journal of Cybersecurity, 15(3), 42-56.*

Suarez Gallegos, M. A. (2024). *Implementation of robotic process automation (RPA) to improve productivity at Covisian Peru SA, Lima.* UWiener Repository.

Suárez Gallegos, M. A. (2024). *Implementation of robotic process automation (RPA) to improve productivity in the company Covisian Perú SA, Lima.* UWiener Repository.

Vega Guevara, W. F. (2021). *Implementation of a Robotic Process Automation (RPA) to improve the logistics management of shipping companies in the company Specialized Reefer Logistics SAC.* UTP Repository.

Zanel, D. (2023). *Master's degree in strategic management of information systems and technologies.* UBA Digital Library.

Zhang, W., & Wei, H. (2022). *Artificial Intelligence and RPA: Advancing Customer Service. AI & Customer Relations Journal, 20(1), 50-58.*

Printed by Books on Demand GmbH, Norderstedt / Germany